THE ANIMAL KINGDOM

走进动物王国

[美] 佚名 著　亢海宏　张颖 译

天津出版传媒集团

百花文艺出版社

图书在版编目（CIP）数据

走进动物王国／（美）佚名著；亢海宏，张颖译. --天津：百花文艺出版社，2016.7（2024.1 重印）
ISBN 978-7-5306-6914-3

Ⅰ.①走… Ⅱ.①佚… ②亢… ③张… Ⅲ.①动物-儿童读物 Ⅳ.①Q95-49

中国版本图书馆 CIP 数据核字（2016）第 167136 号

走进动物王国
ZOUJIN DONGWUWANGGUO

〔美〕佚名 著 亢海宏，张颖 译

| 出版人：薛印胜
| 责任编辑：李 跃
| 美术编辑：郭亚红 封面设计：明轩文化·王 烨
| 出版发行：百花文艺出版社
| 地 址：天津市和平区西康路 35 号 邮编：300051
| 电话传真：+86-22-23332651（发行部）
| +86-22-23332656（总编室）
| +86-22-23332478（邮购部）
| 网 页：http://www.baihuawenyi.com
| 印 刷：河北浩润印刷有限公司
| 开 本：880 毫米×1230 毫米 1/32
| 字 数：113 千字
| 印 张：5.75
| 版 次：2016 年 7 月第 1 版
| 印 次：2024 年 1 月第 2 次印刷
| 定 价：39.80 元

如有印装质量问题，请与河北浩润印刷有限公司联系调换
地址：河北省沧州市肃宁县河北乡韩村东洼开发区 188 号
电话：(0317)3220409

版权所有 侵权必究

〈馆配用书〉

目录

1 The King of Birds 群鸟之王 ... 001

2 Mrs. Zebra and Her Young 斑马夫人和斑马宝宝 ... 004

3 Mrs. Bruin and Her Young 黑熊妈妈和黑熊宝宝 ... 007

4 Little Owls 小猫头鹰 ... 010

5 Aurochs 野牛 ... 013

6 The Kangaroo 袋鼠 ... 016

7 Swans 天鹅 ... 019

8 The Sea Lion 海狮 ... 022

9 Badgers 獾 ... 025

10 The Bird's Nest 鸟巢 ... 028

11　The Chamois 羚羊 ... 031

12　Macaw with Pussy's Bone 与小猫争骨头的金刚鹦鹉 ... 034

13　Members of the Poaching Fraternity 关于偷猎者 ... 037

14　A Cow Working a Pump 会用水泵的母牛 ... 040

15　Carrier Pigeons 信鸽 ... 043

16　Antelope of India 印度羚羊 ... 046

17　Snipe 鹬 ... 049

18　Mrs. Bunny and Family 兔妈妈一家 ... 052

19　The Lynx 猞猁 ... 055

20　The Beaver 海狸 ... 058

21　The Lioness, the Lion and the Cubs 母狮子、雄狮子和小狮子 ... 061

22　A Pet Jack 宠物鱼杰克 ... 064

23　The Swallow's Nest 燕子窝 ... 067

24　Mother-Deer and Baby 鹿妈妈和鹿宝宝 ... 070

25 Whooping Crane 鸣鹤 ... 073

26 The Elk 麋鹿 ... 076

27 Animals Love Toys Too 动物也爱玩具 ... 079

28 The Sucking-Pig 会嘬奶瓶的小猪 ... 082

29 Bell-Ringers 会敲钟的猫 ... 085

30 The Guinea-Pig 豚鼠 ... 088

31 The Argus 阿格斯鸟 ... 090

32 The Clever Fox 聪明的狐狸 ... 093

33 Elephants 大象 ... 096

34 A Wise Dog 聪明的狗 ... 099

35 Our Pet Hedgehog Timothy 我们的宠物刺猬蒂莫西 ... 102

36 The Brave Cockatoo 勇敢的美冠鹦鹉 ... 105

37 Hare Taking the Water 会凫水的兔子 ... 108

38 The Beaks of Birds 鸟嘴 ... 111

39 Blackbirds and Young 大黑鸟和小黑鸟 ... *114*

40 A Useful Pilot 管用的领头羊 ... *117*

41 A bear named Jack 一只名叫杰克的熊 ... *120*

42 A Singular Habit of the Woodcock 丘鹬的奇特习性 ... *123*

43 The Sky-Lark 云雀 ... *127*

44 The Story of a Seal 海豹的故事 ... *130*

45 The Bee 蜜蜂 ... *133*

46 Sheep Dog 牧羊犬 ... *136*

47 The Friendly Terns 有情有义的燕鸥 ... *139*

48 The Otter 水獭 ... *142*

49 The Mastiff 獒 ... *145*

50 The Cunning Wood-Pigeons 狡黠的木鸽 ... *148*

51 Sea Reptiles 海洋爬行动物 ... *151*

52 Swiss Mountain Scenery 瑞士山景 ... *154*

53 Partridge and Young 大鹌鹑和小鹌鹑   ... *157*

54 The Kingfishers' Home 翠鸟的家 ... *160*

55 Rats Carrying Eggs Upstairs 搬鸡蛋上楼的老鼠 ... *163*

56 Heron 苍鹭 ... *166*

57 A Horse Guardian 守护主人的马 ... *169*

58 Battle between a Fox and a Swan 狐狸和天鹅之战 ... *172*

59 Mutual Affection 友情 ... *175*

1

The King of Birds
群鸟之王

As the lion is called the king of beasts, so the eagle is called the king of birds; but except that it is bigger, stronger, and swifter than other birds, there does not seem much reason for the name. It is a mistake to attribute noble or mean qualities to animals or birds, or to think they can do good or bad actions, when they can only do what God has created them to do, and as their instinct teaches.

The most powerful of the eagles is the Golden Eagle, so called because of the rich yellowish–brown bordering to its feathers. It makes its nest in the clefts of the rocky sides of the mountains, and seldom on a tree, unless where one has sprung up in between the clefts, and the tangled roots make a sort of platform. This the eagles cover with sticks, and here they make their house, living in it always, and not only when they lay eggs or have young ones. If there are eagles in the nest, the food is at once carried home to them, and the skinning and eating done at home. Eagles are very attentive to their young, and feed them with great care until they are able to take care of themselves.

正如狮子被尊为群兽之王，鹰也拥有群鸟之王的美誉。鹰之所以有这一美誉，不过是因为和同类相比，它体型更大、身体更壮、行动也更敏捷而已。除此之外，鹰并没有其他特异之处。人们常常把高贵或者邪恶的品性赋予飞禽走兽，以为它们能做好事或者坏事，这是不对的。实际上，它们所做的一切都是出于天性和本能罢了。

　　在所有的鹰中，最厉害的要算是金鹰了。它的每片羽毛上都有一圈金灿灿的黄棕色的边，因此得名金鹰。它一般在高山上的岩缝里筑巢，很少把巢筑在树上，除非这棵树长在山上的岩缝里，并且树根已经盘结成了一个小平台。金鹰用树枝把这个小平台盖起来，筑成窝，不仅在这里下蛋，养育小鹰，还会一直在这里住下去。如果鸟窝里有小鹰，鹰爸爸和鹰妈妈总是在第一时间把吃食送回家，小金鹰蜕皮、吃饭都在窝里进行。鹰爸爸和鹰妈妈对它们的孩子呵护备至，它们会悉心喂养小金鹰直到它们能够自立。

2

Mrs. Zebra and Her Young
斑马夫人和斑马宝宝

Mrs. Zebra, standing with her baby by her side, asks proudly of the lookers-on, "Did you ever see such a likeness?" and certainly mother and child are very much alike, striped all over their bodies, from head to foot, and from nose to tail, with the same regular marks of black. Strong and wild by nature, the zebra family are left very much to themselves, which is a source of great happiness to the mother and child in the picture before us. "No! no! my baby is not going to become as tame as the donkey, or to draw carts and carriages like the horse; it is to have its freedom, and go just where it likes all over these large plains;"—so says Mrs. Zebra, and she means it too, for if anybody took the trouble to go all the way to the hot country of Africa, where Mrs. Zebra is at home, and tried to carry off her baby, they would find their journey a vain one, and that she would kick severely, and perhaps break the legs of the person bold enough to take away her darling.

斑马夫人指着身旁的斑马宝宝，不无骄傲地问路旁的人："你见过比我们更像的母子吗？"的确，这对母子太像了。从头到脚全都整整齐齐地排列着一条一条黑色的斑纹，连鼻子、尾巴上都是。天性桀傲不驯的斑马家族过着无拘无束的生活，你看这幅画上斑马母子看上去多么幸福。 斑马夫人说："不！不！我的孩子绝不会像驴那样俯首听命，也不会像马那样去帮人拉车。它们要有属于自己的自由，可以自由自在地在这一望无际的平原上奔驰。"斑马夫人的这些话可不是闹着玩的。如果有人胆敢不远万里来到炎热的非洲，企图把斑马宝宝从它们的家园带走，那是定难得逞的，因为，斑马妈妈会狠狠地踢他，也许会把这个胆大妄为者的腿踢断。

3

Mrs. Bruin and Her Young
黑熊妈妈和黑熊宝宝

He is the American black bear, who is looking so lively and seemingly inviting the young folks to have a romp, which they will be only too willing to join in. The black bear is of a timid disposition, and seldom attacks man except in self-defense. The female bear is a most affectionate mother, and many stories are related showing her care and love for her young, and her sorrow and mournful cries when any evil befalls them. On one occasion a black bear with her two cubs was pursued across the ice by some armed sailors. At first she urged her cubs to increased speed, but finding her pursuers gaining upon them, she carried, pushed, and pitched them, alternately, forward, until she effected their escape from her pursuers.

黑熊家族生活在美国。它们活泼好动,似乎总在邀请小孩子们和他们一起玩闹,孩子们当然是巴不得呢。黑熊天性温和,如果不是被逼无奈,绝不会攻击人。熊妈妈是天下最慈爱的妈妈,许多故事里的熊妈妈对她的孩子们都是呵护备至、疼爱有加。当孩子遭遇不测的时候,熊妈妈的哭声别提多凄惨了。有一次,一个熊妈妈和两只小熊在冰川上遭到几个持枪水手的追捕。熊妈妈起先是催着熊宝宝快跑,可后来发现捕猎者已经追上来了,于是,熊妈妈连抱带推,有时甚至是抓起熊宝宝往前扔,最后终于脱离了险境。

4

Little Owls
小猫头鹰

Who has not at one time or other of his life read fairy tales and sympathized with stories of enchanted princes and princesses? I once thought of this when a country boy offered me a nest with four of the young of the Little Owl. I put them into a large cage, where they could stare at each other and at my pigeons to their hearts' content.

Let me say that this little owl is a very useful bird, for it keeps mice, bats, beetles, and other creatures in check, which might otherwise multiply too fast. On a spring or summer evening you may hear its plaintive hoot among the apple blossoms of an orchard, or the sheaves of a cornfield. Curiously enough, this simple sound earned the little bird the name of being the harbinger of death, and peasants believed that whenever its cry was heard where sickness was in the family, the patient was sure to die.

在我们的一生中,谁没有读过童话故事?谁没有为那些被施了魔法的公主和王子伤过心?曾经有一个乡村男孩送给我一个鸟窝,窝里还有四只小猫头鹰。那个时候,我不禁产生了这样的感叹。我把小猫头鹰放进一个大笼子里。在那里,它们不仅可以相对而视,还可以细看我的鸽子,想怎么看就怎么看。

告诉你小猫头鹰是益鸟。它捕食老鼠、蝙蝠、甲壳虫还有其他坏家伙,控制它们的数量,以免它们过度繁殖。春夏两季的夜晚,你会听到它们在繁花似锦的苹果园里或者高深茂密的玉米地里哀鸣。这不过是普通的鸟鸣,可奇怪的是,猫头鹰却为此背上了死亡使者的恶名。村民们相信如果家里有病人,一旦听到猫头鹰的叫声,那个生病的人就一定会死。

5

Aurochs[①]
野牛

An Aurochs in blind rage, charging through thick and thin[②], has had a fascination for me as long as I can remember. The true aurochs and this, the European Bison, ceased to exist in the British Isles, except in the Zoological Gardens; but the latter is still found wild in Lithuania, and is also carefully preserved in other parts of Russia, of which the Emperor has a herd. There is much talk about their being untamable—that they will not mix with tame cattle—that tame cows shrink from the aurochs' calves; but does not any cow shrink from any calf not her own? The American Bison, with which you are all pretty familiar, is very similar to the one just mentioned. There have been several attempts made to domesticate the American bison, and have been so far successful. The size and strength of the animal make it probable that if domesticated, it would be of great use.

狂怒中的非洲野牛不顾一切地横冲直撞，这个情景一直让我痴迷。现在，在大不列颠诸岛上已经看不到真正的非洲野牛了，图中的这种欧洲野牛也从大不列颠诸岛上消失了，要看也只能去动物园看；不过，在立陶宛③依然可以找到欧洲野牛的踪影，在俄罗斯的其他一些地方，这种野牛也得到了精心的保护，俄罗斯的沙皇就有一群这样的野牛。人们都说野牛无法驯服，说它们不会和驯牛同居一处，还说驯牛见了小野牛都会远远地躲开；不过，话说回来，所有的牛不都是这样吗？见了别家的小牛犊，都躲得远远的，唯恐避之不及。大家都很熟悉美洲野牛吧？其实这种野牛和我们刚才说的野牛非常相似。有人试图驯化美洲野牛，而且现在看来卓有成效。野牛身强体壮，力大无比，一旦被驯化，将大有用途。

【注】

① Aurochs 和以下出现的 Bison 是居住在不同地方的野牛，在这里统一翻译成野牛。

② through thick and thin：英语习语，意思是"不畏艰难险阻。"

③ 立陶宛现在是一个独立的国家，但是，在历史上曾经几次隶属于俄罗斯。这本书说的就是立陶宛隶属于俄罗斯的时期，因此作者前面提到立陶宛，后面接着说 in other parts of Russian，意思是立陶宛也是俄罗斯的一部分，而且当时的最高统治者是沙皇（Tsar 或 emperor）。

6

The Kangaroo
袋鼠

"Well," said little Herbert Joyce, as he looked over the books of drawings which his cousin had just brought home from Australia, "I never saw anything so extraordinary before in all my life; why here is an animal with three heads, and two of them are very low down, and much smaller than the others." "What do you mean, Herbert?" asked his cousin, who just then came into the room. "There are no three-headed animals—let me see the picture. Oh! no wonder you were puzzled; it does look like a queer creature. That is a kangaroo, and the small heads belong to her children, whom she carries about in a bag formed by a hole in her skin, until they are old enough to walk; and the little things seem very happy there; and sometimes, as their mother moves along over the grass, you may see them nibbling it."

小赫伯特·乔伊斯正在翻看他表哥新近从澳大利亚带回家的画册。"哇,我可从来没见过这么奇异的事情。瞧,这个动物有三个脑袋,两个在下面,而且,这两个脑袋比上面的那个小这么多。"这时,刚好表哥进来,听到他的话,就问:"说什么呢,赫伯特?没有三个脑袋的动物——让我看看。哦!怪不得你觉得奇怪呢,看上去的确很奇怪。这是袋鼠,下面的两个脑袋是两只小袋鼠的。袋鼠妈妈的身上有一个洞,这个洞就是装小袋鼠的袋子。袋鼠妈妈兜着小袋鼠进进出出,直到小袋鼠能走路为止;小家伙们似乎很高兴待在袋子里。有时候,袋鼠妈妈在草地上活动,它们还顺便探出脑袋啃草吃呢。"

7

Swans

天鹅

This beautiful and majestic bird was considered the bird-royal in England, owing to a law of England that when found in a partially wild state on the sea and navigable rivers it belonged to the crown; but of course it is to be found on the ponds and lakes of many a gentleman's estate, and is always prized as a great ornament to the lake. The swan is also very valuable in clearing the ponds of weeds, and makes a most effective clearance, as they eat them before they rise to the surface. The swan affords a pleasing illustration of the love of the mother-bird for its young, and has been known to vanquish a fox who made an attack on its nest—showing that the instinct of motherhood kindles boldness and bravery in the breast of the most timid animals. The nest is generally made on an islet, and composed of reeds and rushes, and when the five or seven large eggs are hatched, the mother may be seen swimming about with the young ones on her back.

在英格兰，人们认为美丽优雅的天鹅是皇室鸟。英格兰有一条法律规定，在海上或通航河流中发现的半野生状态的天鹅都归皇室所有。当然，普通绅士庄园里的池塘和湖泊中也有天鹅，它们点缀其间，成为一道亮丽的风景。天鹅还是很棒的池塘除草工。草还没有长出水面，就被天鹅吃掉了。天鹅又是母爱的典范。曾经有一只狐狸来侵袭天鹅巢，被天鹅打得落花流水。这足以证明母爱的本能可以让最温良的动物胸中燃起无畏的烈焰。天鹅巢一般会筑在礁岩上，里面铺着芦苇和灯芯草。每当天鹅宝宝出世，或五只，或七只，你就会看到天鹅妈妈背着小天鹅在水中游来游去。

8

The Sea Lion
海狮

Although such large and powerful creatures, these sea lions are innocent and playful. See, one of them has reared himself up on his hind legs, if legs they may be called, and is sitting on a chair with his flappers over the back of the chair. It inhabits the eastern shores of Kamchatka, and is in some places extremely abundant, and measuring about fifteen feet in length. It is much addicted to roaring, which, as much as the mane of the old males, has obtained for it the name of the Sea Lion. The old males have a fierce appearance, yet they fly in great haste on the approach of man, but if driven to extremities they will fight desperately; but in captivity they are capable of being tamed, and become very familiar with man. The scientific name of the sea lion is Otary.

别看海狮高大威猛，实际上它们既天真又顽皮。瞧，一头海狮已经撑着后腿站起来了，如果可以把它们叫作腿的话。忽而它又坐到椅子上，还把两只大前掌搭上了椅背。海狮生活在俄罗斯堪察加半岛的东岸。在有些地方，海狮的数量密集，身体长达十五英尺有余。（1英尺=30.48厘米，15英尺大约等于4.57米——译者注。下文中此类换算标准均为译者注）。因为它特别喜欢吼，加上成年雄海狮的脖子上长着鬃毛，所以得名海狮。成年雄海狮长相凶悍。然而，一旦有人靠近，它们会蹿出水面，迅速逃离；不过，如果逼急了，它们也会奋起抵抗。被捕获的海狮是可以驯化的，而且能和人混得很熟。海狮的学名叫海狗(Otary)。

9

Badgers
獾

One day at the Zoological Gardens, I saw the group of Badgers as they are here given. Little do visitors to the gardens take into account how much a wild animal goes through till it has got used to a state of things so opposite to its natural habits. Their wants are attended to as much as possible, but cannot be always met; and so we have here a devoted mother, worn out by the demands of her cubs, and vainly anxious to hide herself from daylight and man's gaze. She has long given up trying to dig or scratch her way out. All she can do is to lean against the wall, ready for a last defence, should anybody come within her prison. She dares not curl up into a ball, like the one cub, and go to sleep; while this little careless imp on her back, happy and trustful, adds to her tiredness by his weight.

一天，我在动物园里看到几只獾，就是图中的这几只。游人们几乎体会不到野生动物要适应与其天性截然相反的一切，需要经历怎样一个痛苦的过程。在动物园，尽管人们尽可能地满足它们的需要，可是，毕竟还会有不如意的地方。不信，你看这位为了孩子甘愿牺牲一切的母亲，已经被几个要这要那的孩子们折磨得精疲力竭了。她多么想躲到暗处，躲开人们的视线，可是却不能。她早已不再掘地挖土，因为她知道无论如何也逃不出去了。现在她所能做的就是身子抵住墙，随时准备对任何进入监牢的人发起最后的反击。她可不敢像小獾那样蜷作一团，安然入睡。然而，这个不懂事的小淘气趴在妈妈的背上却那么快乐，那么安逸；他哪里知道，妈妈早已不堪重负了。

10

The Bird's Nest
鸟巢

Her little nest, so soft and warm,
God teaches her to make it;
I would not dare to do her harm,
I would not dare to take it.

How curious is the structure of the nest of the Bullfinch (红腹灰雀) or Chaffinch(苍头燕雀)! The inside of it is lined with cotton and fine silken threads; and the outside cannot be sufficiently admired, though it is composed only of various kinds of fine moss. The color of these mosses, resembling that of the bark of the tree in which the nest is built, proves that the bird intended it should not be easily discovered. In some nests, hair, wool, and rushes are cleverly interwoven. In others, the parts are firmly fastened by a thread, which the bird makes of hemp, wool, hair, or, more commonly, of spiders' webs. Other birds—as, for instance, the blackbird and the lapwing—after they have constructed their nests, plaster the inside with mortar; they then stick upon it, while quite wet, some wool or moss to give warmth; but all alike construct their nests so as to add to their security.

她的小窝又软又暖和，
上帝教她如何把巢筑；
我可不敢把她来伤害，
我也不敢把她带回家。

无论是红腹灰雀，还是苍头燕雀，它们的鸟巢构造实在是太奇特了！里面密密麻麻布满了棉线和细丝线；外面只不过是各种各样的细苔藓，但是其精美程度简直让人叹为观止。鸟巢筑在树上，这些苔藓的颜色接近树皮的颜色，显然，鸟儿们可不想轻易暴露自己的窝。有些鸟巢里，发丝、羊毛和灯芯草被巧妙地编织在一起。还有一些鸟巢的筑巢材料是用一根线紧紧地连在一起的，这根线也许是麻线，也许是羊毛线，也许是发丝，更常见的则是蜘蛛网。还有些鸟，比如山鸟、田凫，筑完巢之后，会在巢的里面抹上一层泥，然后，趁着泥还没干的时候，沾一些羊毛或者苔藓上去，这样利于保温。不过，归根结底，鸟儿们筑巢都是为了安全。

11

The Chamois
羚羊

The chamois are indeed high-born, for among the high mountain peaks, where the eternal snow rests and the Alpine roses bloom, there they make their home! There they spring up over the snowy slopes to those heights to which man cannot climb. They rest upon the glittering ice, the snow does not blind them, neither does it cool their hot blood. Carelessly they stride across the snowed-over crevices, and when the terrible storms, at which men are so alarmed, hurl down rocks and avalanches from the summits, the Chamois do not fear them. They find their way safely through the thickest mist and darkest clouds. Agile and light-footed, gentle and peaceable, proud and courageous, they lead a happy life among the mountains, as long as man does not molest them.

羚羊出身可真"高"呀！因为它们在高耸入云的山峰之间安家落户，那里终年积雪，阿尔卑斯高山玫瑰漫山遍野。在那里，它们能越过冰雪覆盖的山坡跳到人们无法企及的峭壁上。晶莹的冰面上，茫茫白雪无法刺瞎羚羊的双眼，也无法冷却它们的热血。积雪覆盖下的裂缝危机四伏，可它们却能漫不经心地一一大踏步跨过。暴风雪来临时，山峰上滚落的石块和雪崩让人类心惊胆战，可是羚羊们却毫不畏惧。即使迷雾重重、乌云笼罩，它们也不会迷失方向。它们轻盈敏捷、从容淡定、骄傲勇敢，只要人类不来骚扰，它们在崇山峻岭间过得很快乐。

12

Macaw with Pussy's Bone
与小猫争骨头的金刚鹦鹉

Jacko is a bird called a Macaw, and has fine feathers—scarlet and yellow and blue. Jacko can talk a little. He says, "Come along, Jacko, come along;" and when you come, as soon as he thinks you near enough, he pecks at you with his great beak. When he is in a good temper he will say, "Poor, poor! " He will sit upon the ivy all the morning and talk to himself, and he will call the gardener, and he will cough and sneeze, and crow and cackle, in a very funny manner. If Jacko sees sparrows picking up a few crumbs, he will rush up, sweeping his great wings along the ground, and take their meal for himself. If he sees poor Pussy picking a bone, he takes great delight in creeping down from his ivy, helping himself down with beak and claws, and at a sight of Jacko's approach Pussy darts away, leaving the bone in Jacko's possession. Pussy, of course, does not like this, but stands at a respectable distance, and with curved back and flashing eyes shows her indignation at Jacko. Presently Jacko retires to the ivy and Pussy resumes her feast.

咕咕是一只金刚鹦鹉，一身斑斓艳丽的羽毛——猩红、金黄，还有天蓝。咕咕会说一点人语，它说："过来，咕咕，过来。"如果你真的过去，还没到跟前，它就会伸出大大的鸟喙来啄你。脾气好的时候，它会说"好可怜，好可怜！"它能整个上午都待在常春藤上，一会儿自言自语，一会儿呼喊花匠，一会儿咳嗽打喷嚏，一会儿又叽咕咕咕地叫个不停，那神态简直滑稽极了。看到麻雀啄食地上的面包渣，它就张开两只大翅膀横扫过来，把这顿美餐据为己有。看到可怜的小猫啃骨头，它会满心欢喜地溜下常春藤，连啄带抓地不请自吃起来。小猫看到咕咕，马上躲在一边，眼睁睁看着骨头落入咕咕之手。当然，小猫并不甘心，它站在不远处，弓起背，瞪起眼，怒目而视。没多久，咕咕又飞回到常春藤上，小猫也接着享受它的盛宴。

13

Members of the Poaching Fraternity
关于偷猎者

Among the various wild animals which inhabit the earth, it is difficult to decide which are really friendly and which are really hostile to man's interests. The actual fact appears to be that there is neither hostility nor friendship. If farmers and gardeners kill off too many birds, nature revenges herself by sending a plague of insects which the small birds, if alive, would have eaten. Gamekeepers ruthlessly shoot hawks and kites, or snare stoats and polecats, with the result that their game grows up too thick for its feeding ground, sickly specimens are allowed to linger on, and a destructive murrain follows. The rook, no doubt, is fond of eggs; but nevertheless he does the farmer good service when he devours the grubs which are turned up by the plow; and as the salmon disease, which of late has proved so destructive, is attributed by the best authorities to overcrowding, that glossy-coated fisherman, the otter, is really a benefactor to the followers of Izaak Walton's gentle craft.

地球上生活着各种各样的野生动物,哪些是我们的朋友,哪些是我们的敌人?这很难说。实际上,根本就没有什么敌人和朋友。如果农夫和园丁杀死太多的鸟,大自然就会报复,让那些本来可以被鸟吃掉的虫子大肆泛滥。猎场看守人无情地射杀鹰和鸢这样的猛禽,或者诱捕白鼬和臭鼬,它们的猎物就过度繁殖,其觅食地也因此而变得不堪重负,羸弱的动物也得不到及时的淘汰,结果毁灭性的瘟疫就接踵而至。毫无疑问,秃鼻鸦喜欢偷吃鸟蛋,但是,农民犁地的时候,它们也会吃掉从地下翻出来的虫子,这倒帮了农民的大忙;另外,根据最权威的解释,刚刚被证明极具毁灭性的鲑鱼病是由于鲑鱼数量过多引起的,由此看来,穿着溜滑外套的捕鱼高手水獭可真是那些艾萨克·沃尔顿垂钓技巧信徒们的大恩人呀!

【注】

Izzaak Walton 艾萨克·沃尔顿,17世纪英国的一位垂钓爱好者,写了著名的《垂钓大全》(*The Compleat Angler*),该书介绍了钓鱼技巧、场地选取、鱼饵选取等,也是一本语言优美的散文书。The followers of Izaak Walton's gentle craft 指那些垂钓爱好者。这句话的意思是说,对于那些喜欢钓鱼的人来说,捕鱼高手水獭能够让鱼类保持一定数量平衡,而避免了鱼类疾病的盛行,从而也保护了鱼类,使喜欢钓鱼的人能有鱼可钓,因此,他们应该感谢水獭。

14

A Cow Working a Pump
会用水泵的母牛

My informant writes me as follows: "We have a wonderful cow here—about ten years old, and very clever at opening gates and breaking fences. There is a pump about three feet high in the center of the field, near my house, over a trough, which is, or ought to be, filled daily. It was on a hot day, when my man had omitted to pump the trough full, that the cow was first observed to help herself: the way in which she managed to pump was by pushing the handle up with her head and then forcing it down with her horns. Very little elevation of the handle is required to get water, and she would work it for five minutes together, and sometimes drank from the spout, and sometimes from the trough."

有人给我写信爆料："我们这里有一头了不起的母牛——大约十岁,很聪明,会开门,还会破篱而入。我家附近的那块田中央,有一个差不多三英尺高的水泵(3英尺=91.44厘米),泵下面是一个水槽,每天都得给这个水槽加满水。有一天,天很热,我雇的人忘了给水槽加水了,于是,这个母牛竟然自力更生,自己用水泵打起了水。这可是破天荒第一次。她汲水的方法如下:先用头把水泵的摇把顶上去,然后再用犄角把它压下来。那个摇把只要往上抬一点点,就能把水打上来,她前后只要花五分钟时间。有时候她抬头去喝泵口的水,有时候又低头去喝水槽里的水。"

15

Carrier Pigeons

信鸽

The carrier pigeon is remarkable for the degree in which it possesses the instinct and power of returning from a distance to its accustomed home. In Eastern countries it is the practice to bathe the pigeon's feet in vinegar to keep them cool, and to prevent it from alighting in quest of water, by which the letter might sustain injury.

Pigeons intended for this use must be brought from the place to which they are to return, within a short period, and must be kept in the dark and without food for at least eight hours before being let loose. The carrier pigeon was of great service during the siege of Paris in 1871, and conveyed many important messages. It goes through the air at the rate of thirty miles an hour, but has been known to fly even faster.

信鸽具有超常的直觉和力量,能越过千山万水返回故土。在东方国度里,人们常常用醋给鸽子洗脚,为的是让鸽子脚部保持凉爽,以防它为了找水而降落,从而危及信件的安全。

肩负送信使命的鸽子,短期内必须到达目的地。放飞前,必须在黑暗中待上至少八小时,其间不能进食。在1871年的巴黎围攻战中,信鸽传递了许多重要的消息,发挥了巨大的作用。信鸽的飞行速度是每小时三十迈(约48公里/每小时),但是据说远不止这个速度!

16

Antelope of India
印度羚羊

The antelope of India, roams over the open and rocky plains of that immense country. It is distinguished from the rest of its family by the beauty and singular shape of its horns, which are annulated or ringed, and spirally convoluted or curved together, making two or more turns, according to the age of the animal. The fakirs and dervishes of India, who are enjoined by their religion from carrying swords, frequently wear at their girdles the polished horns of the antelope instead of the usual military arm. This antelope is one of the fleetest-footed of its family, and its leap is something wonderful. It is not uncommon for it to vault to the height of twelve or thirteen feet, passing over ten or twelve yards at a single bound. In color it is almost black on the upper part of the body, and light-colored beneath. When full grown, it is about the size of our common deer.

印度羚羊徜徉在印度那一望无际、岩石嶙峋的平原上。与其他羚羊不同的是，印度羚羊的犄角成环状向上盘旋、扭曲，中间视年龄不同而弯成不同个数的弯，看上去美丽而又奇特。在印度，由于宗教信仰不允许苦行僧们佩戴利剑，所以，他们经常把打磨光滑的羚羊角佩戴在腰间，作为平时的防身武器。印度羚羊的腿脚在羚羊家族中最灵活，弹跳力也棒极了。对它来说，一跳跳个十二三英尺（约三四米）高，十到十二码（1 码＝0.9144 米，10 码到 12 码约合 9 米到 11 米）远简直是稀松平常。它的肤色，上半身几乎是黑的，下半身则是浅色。成年羚羊的大小和一只普通的鹿差不多。

17

Snipe
鹬

These birds frequent swampy woods, marshes, morasses, and the borders of rivers. Their usual time for seeking their food is early in the morning and during the twilight of the evening. They subsist principally upon insects and worms; for these they search among the decayed leaves, and probe the mud and ooze with their lengthened bills. When alarmed, they generally lie close to the ground, or among the grass, or, suddenly starting on the wing, escape by flight, which is short but elevated, rapid, and irregular. The eggs, which are four in number, are deposited on the ground. In the snipe, and all its immediate allies, the bill is thickened, soft, and very tender at its extremity; so that this part, which is richly supplied with nerves, serves as a delicate organ of touch, and is used for searching in the soft ground for the insects and worms that constitute the food of these birds.

鹬出没于泥泞潮湿的树林、沼泽、湿地和河边。它们一般在清晨或者黄昏时分出来觅食，主要吃昆虫和蠕虫。为了寻找这些食物，它们伸着长长的嘴巴在枯枝烂叶和淤泥沼泽中东翻西探。如果受到惊吓，一般情况下，鹬会俯卧在地上，或者躲在草丛里，有时候，也会突然张开翅膀，冲上天空逃走。它不会飞很远，但会飞很高，整个过程敏捷从容。鹬一窝下四个蛋，通常产在地上。鹬的嘴巴和它所有近亲的嘴巴一样，顶部比较厚，而且比较软，那里神经分布密集，所以非常敏感，是个精巧的触觉器。这些鸟儿就用它来翻找软土中的昆虫和蠕虫等美食。

【注】
　　这里介绍的"鹬"，就是我国的成语故事"鹬蚌相争，渔翁得利"中的那个"鹬"。

18

Mrs. Bunny and Family
兔妈妈一家

This wild rabbit has been startled by some noise, and the next moment she may be scampering away to her burrow, with the little bunnies, at the top of their speed, and crouch there until all is quiet again. Rabbits usually select, if possible, a sandy soil overgrown with furze, in which to make their burrows, as such a soil is easily removed, and the dense prickly furze hides their retreat, whilst it affords them a wholesome and never-failing food. These furze bushes are constantly eaten down, as far as the rabbits can reach standing on their hind legs, and consequently present the appearance of a solid mass with the surface even and rounded. These animals retire into their burrows by day to rest, and come out only in the twilight to obtain food.

这些野兔呀，很容易受到惊吓。瞧，稍有响动，兔妈妈就会带着她的小宝宝们拼命奔到洞里，蜷缩起身子躲起来，直到一切归于平静。一般情况下，兔子们把窝建在长满荆豆的沙地里，因为这样的地容易挖。而且，荆豆上长满了密密麻麻的刺，不但能掩护它们逃命，还能源源不断地为它们提供有益健康的食物。那一丛丛的荆豆，只要是兔子能够凭借后腿支撑够得着的，总是被啃得齐刷刷的，露出圆圆的一大块。兔子这种动物白天躲在洞里休息，晚上才出来找吃的。

19

The Lynx
猞猁

The body of the lynx, beautifully spotted with black and brown rings, is more solid and hardy than that of the wild cat. His ears are longer, his tail is shorter, his great eyes light up like bright flames; and since he prowls about chiefly at night, he is thought to have very keen sight. For this reason, when we wish to say that a person can see very clearly or can look beyond the outward appearance of things, we call him *lynx-eyed*. Like all cats, the lynx possesses in his mustache a very correct power of feeling. This, with the sense of hearing and sight, guides him in all his expeditions. The lynx in the picture is in the act of springing upon a timid hare. Although he can measure twenty paces in a jump, I think for once he has made a misstep, and the dear little creature with one more bound will be safe. One very remarkable fact about these animals is this: if there are several together, and one starts over the snow in pursuit of booty, all the others will follow in exactly the same tracks, so that it will look as if but one lynx had passed over the snow-covered earth.

猞猁身上点缀着黑色和棕色小圈，非常漂亮，而且，身体比野猫还结实。它耳朵较长，尾巴稍短，目光炯炯，像两团燃烧的火焰。因为猞猁主要是晚上出来活动，所以，大家都认为它的视力特别好。也正因如此，如果一个人的眼神很好，或者能透过事物的表面看清其本质，我们就会说他长了"猞猁眼"。和所有的猫科动物一样，猞猁的胡子具有很强的感觉能力，超常的感觉加上超常的听觉和视觉，使猞猁在所有的行动中都游刃有余。图片中的这只猞猁正在扑向一只受惊的野兔。尽管它一下就能跳二十步，可是它可能不小心失了脚，所以那只可怜的小野兔再蹦一下就可以躲过一劫了。这种动物还有一个让人称奇的特点：好几只猞猁在一起时，如果其中一只去追猎物，其余猞猁也会一步不差地沿着它的足迹去追。这样，从雪地上留下的脚印来看，跑过去的就像是一只猞猁，而不是好几只。

20

The Beaver
海狸

This industrious animal is generally found in Canada and the northern portions of the United States, where it makes its home on the banks of the rivers and lakes. Here they assemble in hundreds to assist each other in the construction of their dams, and in the building of their houses, which are put together with a considerable amount of engineering skill. The materials used in building the dams are wood, stones, and mud, which they collect themselves for that purpose, and after finishing the dam, or winter storehouse, they collect their stores for the winter's use, and then make a connection with their houses in the banks. Their skins are valuable in making fine hats, and their flesh is much relished by the hunters. The beaver is an interesting animal in many respects, and the expression "busy as a beaver" is borne out by its habits.

我们可以在加拿大和美国北部看到这种勤劳的动物，它们在河流和湖泊的堤岸上安营扎寨。筑堤建房时，它们会召集上百号的伙伴互相帮忙，它们修筑的堤坝和建造的房子技术含量相当高。建造堤坝的材料有木头、石块和泥浆，都是它们为筑堤建房而亲自搜集的。堤坝是它们的冬储仓库，完工之后，它们就开始为过冬做储备，然后，用隧道把建在岸上的房子连起来。海狸皮是制作精致帽子的珍贵材料，海狸肉也让捕猎者们津津乐道。在很多方面，海狸是一种有趣的动物，人们常说"像海狸一样忙"，就是因为海狸总是忙个不停。

21

The Lioness, the Lion and the Cubs
母狮子、雄狮子和小狮子

The lioness is much smaller than the lion, and her form is more slender and graceful. She is devoid of the mane of her lord and master, and has four or five cubs at a birth, which are all born blind. The young lions are at first obscurely striped and spotted. They mew like cats, and are as playful as kittens. As they get older, the uniform color is gradually assumed. The mane appears in the males at the end of ten or twelve months, and at the age of eighteen months it is very considerably developed, and they begin to roar. Both in nature and in a state of captivity the lioness is very savage as soon as she becomes a mother, and the lion himself is then most to be dreaded, as he will then brave almost any risk for the sake of his lioness and family.

母狮子个头比雄狮子小得多,身材也比雄狮子苗条优雅得多。她身上没有一家之主雄狮子那样的鬃毛。母狮子一次能产四五只幼仔,刚出生的小狮子什么也看不见。起初,小狮子身上的条纹和斑点不太清楚,它们像猫一样咪咪叫,像小猫一样顽皮。再大一点,它们身上的花纹逐渐清晰起来。到了十个月或者十二个月的时候,小雄狮子的身上开始长鬃毛。十八个月大的时候,小狮子就发育得差不多了,并且开始吼叫。无论是在野生状态下,还是在圈养状态下,母狮子一旦做了母亲,性情就变得非常暴烈;而此时的雄狮子也最可怕,因为,为了老婆和孩子,它会不顾一切。

22

A Pet Jack
宠物鱼杰克

The first fish I ever saw in an aquarium, twenty years ago, was a "Jack," as he is called when young, or a "Pike," when he grows older; and ever since then I have contrived to have a pet fish, and this, drawn from life by Mr. Harrison Weir, is an accurate portrait of the one I now possess in the Crystal Palace Aquarium.

There he is, just as he steals round the corner of a bit of rock. He is glaring at a minnow, at which he is taking most accurate aim; he hardly seems to move, but yet he does by a very trifling motion of the edge of his back fin—sometimes resting a little on the tips of his two foremost fins, as they touch the ground, carefully calculating his distance; and then, at the very moment when the minnow has got into a position which leaves a space of clear water in front, so that Mr. Jack shall not hurt his nose against any hard substance when he gets carried on by the violence of his rush, he darts at the

minnow with the speed of Shakespeare's Puck:—
"I go, I go! look, how I go!
Swifter than arrow from the Tartar's bow."

二十年前，我在水族馆里见到的第一条鱼是一条"杰克鱼"。它小时候，大家都这样叫它；长大了，大家就叫它"梭子鱼"。从那时起，我就想有一条宠物鱼，而这幅画里的鱼就是我的，它现在生活在水晶宫水族馆里。这幅逼真的肖像画是哈里森·韦尔照着真鱼画的。

你瞧它，从一块石头后面悄悄地探出身子，紧紧盯着一条桃花鱼，瞄准，随时准备发起致命的进攻。看上去它几乎一动不动，其实它在动，它的背鳍边缘在微微摆动——有时候，它的两只前鳍尖触到鱼箱底，稍事休息，仔细测算着前面的距离。好！机会来了！此时桃花鱼所处位置的前方刚好有一片开阔的水域，"杰克先生"此时若冲过去，不会因为用力过猛撞上前面的硬物而碰坏鼻子，于是，说时迟那时快，它以莎士比亚笔下小精灵帕克的速度冲了过去：

"冲呀，冲呀，你看我怎么冲！
比鞑靼人射出的箭还要快。"

23

The Swallow's Nest
燕子窝

Often in former years the twitter of the birds glittering in the morning sun was the first sound that met my ear during the wakeful hours which frequently accompany illness after the worst crisis has passed, and you are recovering by degrees. The gutters ran beneath my bedroom windows, and I could see the steel-blue backs of the swallows as they sat on the rims of the gutter, twisting their little heads, opening their yellow-lined beaks, singing to their hearts´ content. Whole families would perch there together, or the young would rest in rows of four or five, according to the nest-broods of each. How delightful to see them fed by their agile parents! How tantalizing to have them almost within reach of my hands, yet not to be able to catch them or give them a kiss, as they would cower in my hollow hands if I only could have got them in there!

前几年，我大病初愈，辗转难眠。每天清晨，耳旁响起的第一个声音就是灿烂阳光下鸟儿们叽叽喳喳的鸣叫，于是身体也渐渐恢复。我的卧室窗下是排水槽，当燕子们落在水槽边上的时候，我可以看到它们灰蓝色的后背。它们扭着小脑袋，张开黄灿灿的小嘴巴，尽情歌唱。它们会全家栖息在一起，有时候，小燕子们四个或五个排成一排，这要看一窝有几只小燕子了。看着它们的爸爸妈妈敏捷麻利地给小燕子喂食，简直太开心了！不过，看着眼前触手可得的燕子，不能抓，也不能亲，别提多难受了！唉！如果我真的把它们捧在手掌里，它们一定会吓得缩成一团的。

24

Mother-Deer and Baby
鹿妈妈和鹿宝宝

Something has startled them, as they fed securely enough, one would think, on the grass at the foot of the rocks; and if we could only get a little nearer, this is what we should hear the mother-deer saying to her baby: "My child, I am sure there is danger about; look out and tell me if you see the slightest movement on the hill yonder, or if I see it first, I will give you the signal, and you must follow me, and run for your very life." And the baby, with cocked ears and glistening eyes, promises to do as it is told. But after all it will probably prove a false alarm, for this is not the time of year for deerstalking; and I dare say the noise they heard was made by a party of people coming up the valley below to see the waterfall, which is famous in the neighborhood.

瞧这幅画！鹿妈妈和鹿宝宝正在岩石下悠然自得地吃着草，突然它们听到了什么动静……假如我们走近一些，就会听到鹿妈妈在对鹿宝宝说："孩子，有危险。当心山那边的动静，哪怕是一点点动静也要告诉我。要是我先发现了危险，就发信号给你，你可千万要跟着我，拼命逃呀。"鹿宝宝竖着耳朵，眨巴着眼睛，答应妈妈一定听话。然而，这很可能是虚惊一场，毕竟现在不是猎鹿的季节。我敢说，它们听到的动静是下面山谷里的一群游人发出的，因为附近有个瀑布，很有名，他们是赶来看瀑布的。

25

Whooping Crane
鸣鹤

The Whooping Crane is much larger than the common crane, which it otherwise much resembles except in color; its plumage, in its adult state, is pure white, the tips of the wings black. He spends the winter in the southern parts of North America, and in summer migrates far northwards. The crane feeds on roots, seeds, etc., as well as on reptiles, worms, insects, and on some of the smaller quadrupeds. They journey in flocks from fifty to a hundred, and rise to an immense height in the air, uttering their loud harsh cries, and occasionally alighting to seek food in fields or marshes; and when they descend on a field they do sad havoc to the crops, several doing sentinel duty while the majority are feeding. In general it is a very peaceful bird, both in its own society and those of the forest.

和普通的鹤相比，鸣鹤要大得多，颜色也不一样，除此之外，它们看起来倒是很像。长大后的鸣鹤一身纯白，只有翅膀顶端是黑的。它们在北美南部过冬，夏天，则向北迁徙到很远的地方。鸣鹤以植物的根、种子等为食，也吃爬虫、蠕虫、昆虫和其他一些小的四足动物。它们成群结伴而行，每次五十到一百只不等。它们飞得很高，一边飞还一边沙哑地高声鸣叫。偶尔落在田野里或湿地里找食吃，这时就会给庄稼带来灭顶之灾。大部队找食吃的时候，还有几只得站岗放哨。总体来说，无论在自己的小圈子内，还是在整个大森林里，鸣鹤都算是一种比较温和的鸟。

26

The Elk
麋鹿

This is the largest existing species of the deer family, and is a native of the northern parts of Europe, Asia, and Africa. It grows to be six feet high and twelve hundred pounds in weight. They are very rare in Europe and America, but at one time they extended as far south as the Ohio River. They love the woods and marshy places, and live off of the branches of trees, being unable to eat grass unless they get upon their knees. They are very timid, and easily approached by the hunter, but should a dog come in the way, one stroke from an elk's foot will kill it. Many of the parents of our little friends in Maine and Canada are, no doubt, familiar with the elk and its habits.

鹿家族现存的各种鹿中,麋鹿是最大的一种。它生活在欧洲、亚洲和非洲的北部,能长到六英尺(约一米八三)高,一千二百磅(1磅＝0.4536千克,一千二百磅约合五百四十千克)重。在现在的欧洲和美国,麋鹿是稀有动物,可是,它的足迹曾经向南最远到过俄亥俄河。麋鹿喜欢树林和沼泽。由于麋鹿必须双膝跪地才能吃到草,所以它们以吃树叶为生。麋鹿很温顺,容易成为猎人的囊中之物。但是,如果狗挡了它的路,麋鹿只消一脚,就会要了狗的命。缅因州和加拿大的小朋友们,你们的爸爸妈妈肯定很熟悉麋鹿,也很了解它们的习性哟!

27

Animals Love Toys Too
动物也爱玩具

The "Daily News" says: "Our readers have often doubtless observed appeals in the papers for toys for sick children. We hear that a naturalist who feels much for animals is struck with the cruelty of leaving the creatures at the 'Zoo' without anything to play with. This gentleman had in his possession a young otter, for whom he made a wooden ball, to the extreme delight of his pet, who used to divert his simple instinct with it for whole hours at a stretch. Following up the idea, the same gentleman presented the elephants and rhinoceroses in the Zoological Gardens with globes for diversion suited to their sizes, but it seems the elephants took to playing ball so furiously, that 'there was danger of their houses being swept down altogether; so they were forbidden to use them indoors.' The polar bear was given a toy which, we are told, 'amuses him immensely.'"

《每日新闻》报道说:"毫无疑问,我们的读者朋友们经常在报纸上看到一些文章,呼吁给生病的孩子捐赠玩具。听说有一个热爱动物的博物学家觉得把动物关在'动物园'里却不给它们玩具玩简直太不人道了。这位先生有一只小水獭,于是他给小水獭做了一个木球,这可把小水獭给乐坏了,一玩就是好几个小时。由此这位先生受到启发,又给动物园里的大象和犀牛带去了大小合适的球形玩具。可是大象玩球玩疯了,'都快把房子掀起来了,所以,后来就不许它们在室内玩球了。'北极熊也得到了一个玩具,据说'这个玩具让它兴奋不已。'"

28

The Sucking-Pig
会喂奶瓶的小猪

The other day our children came home delighted at having seen a little pig drinking out of a bottle, just like a baby. I went to see it, and I was introduced to its owner, who lived in a cottage, the principal room of which was painted light blue. A good-natured old woman was there with her two orphan grand children. The red tiles of the cottage floor were enlivened by a gray-and-white cat, and a shiny-skinned little pig, of about a month old, which was fed out of a feeding-bottle. This was the hero of the place.

The little pig is grateful for good treatment, and as capable of attachment as a horse or a dog. The pig is intelligent, and it can be taught tricks. Performing pigs are often the attractions of country fairs. I have seen pigs in the poor neighborhoods of London follow their masters through noisy streets, and into busy public houses, where they laid down at their masters' feet like a dog.

有一天,孩子们回家,兴奋地告诉我说,他们看见一只小猪会像婴儿一样喂奶瓶。于是我出门去看个究竟。经人指引,我来到小猪的主人家。那是一个村舍,正房被刷成了淡蓝色,里面住着一位慈祥的老婆婆和她两个失去了爹妈的孙子。在红砖铺成的地板上,我看到一只灰白色的猫和一只皮肤光亮的小猪,它们的存在让小屋熠熠生辉。小猪大约有一个月大,主人正在用奶瓶给它喂食。这就是我要拜访的那只小猪了。

得到悉心照顾的小猪,也会像马和狗那样对主人忠心耿耿、形影相随。猪很聪明,能学会一些小把戏。会表演把戏的猪在集市上常常引人注目。我在伦敦的乡下就看到过几只猪跟着它们的主人穿过嘈杂的街道,进了酒吧,然后像狗一样卧在主人的脚边。

29

Bell-Ringers
会敲钟的猫

When I was a child, my father took me to see some feats performed by some traveling cats. They were called "the bell–ringers," and were respectively named Jet, Blanche, Tom, Mop, and Tib. Five bells were hung at regular intervals on a round hoop erected on a sort of stage. A rope was attached to each bell after the manner of church bells. At a given signal from their master, they all sprang to their feet, and at a second signal, each advanced to the ropes, and standing on their hind feet, stuck their front claws firmly into the ropes, which were in that part covered with worsted, or something of the kind, so as to give the claws a firmer hold. There was a moment's pause—then No. 1 pulled his or her rope, and so sounded the largest bell; No. 2 followed, then No. 3, and so on, till a regular peal was rung with almost as much precision and spirit as though it were human hands instead of cats' claws that effected it.

小时候,爸爸带我去看游走江湖的猫表演节目。这些猫被叫作"敲钟人",它们的名字分别是杰特、布朗、汤姆、忒布。台上立着一个铁环,上面均匀地挂着五个铃铛,每个铃铛上都系着一根绳子,就像教堂的钟那样。主人发出第一声信号,五只猫都跳起身,发出第二次信号,它们就跑到绳子跟前,后脚着地立起来,前爪紧紧地扣住绳子,抓手的那段绳子上缠着布或其他诸如此类的东西,这样就能抓紧了。稍事片刻,一号猫开始拉绳子,最大的那个铃铛就响起来了。接下来是二号猫,然后是三号,如此轮下去,直到铃铛有节奏地响起来,其准确与响亮的程度,如同出自人手,而不是猫爪。

30

The Guinea-Pig
豚鼠

The guinea-pig is a native of South America, and is remarkable for the beauty and variety of its colors, and the neatness of its appearance. These little pets are very careful in keeping themselves and their offspring neat and tidy, and may be frequently seen smoothing and dressing their fur, somewhat in the manner of a cat. After having smoothed and dressed each other's fur, both turn their attention to their young, from whose coats they remove the smallest speck of dirt, at the same time trying to keep their hair smooth and unruffled. The Guinea-pig feeds on bread, grain, fruit, vegetables, tea leaves, and especially garden parsley, to which it is very partial. It generally gives birth to seven and eight young at a time, and they very soon are able to take care of themselves.

豚鼠土生土长在南美。它们身上有好几种漂亮的颜色，看上去又干净又整洁，分外引人注目。这些小东西特别在乎外表，总是把自己和孩子们收拾得干干净净、整整齐齐的。它们经常像猫那样梳理自己的皮毛。先是豚鼠夫妻俩互相梳理，然后一起为孩子们梳。它们把孩子们身上的脏东西弄掉，哪怕是最小的渣滓也不放过。在这个过程中，还要保持自己身上的毛发纹丝不乱。豚鼠吃面包、谷粒、水果、蔬菜、茶叶，尤其爱吃菜园里的香菜。豚鼠通常一次能产七到八只小豚鼠，而且，小豚鼠很快就能自己照顾自己。

31

The Argus[①]
阿格斯鸟

The Argus is a bird with magnificent plumage; it inhabits the forests of Java and Sumatra, and takes its place beside the pheasant, from which it only differs in being unprovided with spurs, and by the extraordinary development of the secondary feathers of the wings in the male. The tail is large and round, and the two middle feathers are extremely long and quite straight. When paraded, as it struts round the female, spreading its wings and tail, this bird presents to the dazzled eye of the spectator two splendid bronze-colored fans, upon which is sprinkled a profusion of bright marks much resembling eyes. It owes its name of Argus to these spots.

阿格斯鸟身着华丽的羽毛,生活在爪哇岛和苏门答腊岛上的森林里。这种鸟和雉鸡差不多,唯一的不同就是阿格斯鸟天生没有肉距,而且雄鸟的次生羽翼超级发达,尾巴又大又厚,尤其是中间的两根羽毛又长又直。每当雄鸟张开翅膀和尾巴,围着雌鸟神气活现地走来走去炫耀自己的时候,就会展现出两个精美绝伦的古铜色扇子,扇子上撒满了像眼睛一样的亮圆圈,令人眼花缭乱。正因如此,它才得名阿格斯。

【注】
① Argus 是希腊神话中有一百只眼睛的巨人,非常机警,睡觉时从不把眼睛全部闭上。

32

The Clever Fox
聪明的狐狸

One summer's day on the banks of the river Tweed, in Scotland, a fox sat watching a brood of wild ducks feeding in the river. Presently a branch of a fir tree floated in their midst, which caused them to rise in the air, and after circling round for some time, they again settled down on their feeding ground. At short intervals this was repeated, the branch floating from the same direction, until the ducks took no further notice of it than allowing it to pass by. Mr. Reynard noticed this; so he got a larger branch than the others, and crouching down among the leaves, got afloat, and coming to the ducks, who took no notice of the branch, he seized two of the ducks, and then allowed himself to be floated to the other side, where, we suppose, he had a repast.

有一年夏天，一只狐狸蹲在苏格兰崔德河的河岸上，看着一群野鸥在河里觅食。很快，河上飘过来一根杉树枝，吓得野鸥立刻飞了起来，盘旋了一会儿之后，它们又落下来觅食。由于树枝不断地从相同的方向漂来，这一幕时隔不久就要重演一次。直到后来，野鸥不再理会这些树枝，任由它们漂过。狐狸先生将这一切看在眼里。于是，它找了一根大树枝，趴了上去，用树叶遮住自己，下了水，漂到已经对树枝毫不在意的野鸥跟前，逮到了两只野鸥，然后又顺水漂到了河对岸。在那里，它一定是美美地吃了一顿。

33

Elephants
大象

See this monster of the forest uprooting trees, as a test of its strength before entering on a fight with one of its companions, which is often a bitter struggle for supremacy. There are two species of Elephants, the Indian and African; the ears of the latter are much larger than the Indian, covering the whole shoulder, and descending on the legs. Elephants live in herds, and each herd has a leader—generally the largest and most powerful animal—who exercises much control over the herd, directing its movements, and giving the signal in the case of danger. The trunk of the Elephant is of great service to it, and is a wonderful combination of muscle; Curier, the famous Naturalist, stating that there is not far short of 40,000 muscles, having distinct action, and so giving it an acute sense of touch and smell—so much so, that it can pick up a pin, or pluck the smallest leaf. The Elephant is generally about ten feet high, and sometimes reaches to twelve feet, and lives to the age of seventy or eighty years.

瞧，这头森林巨兽正在拼命地将一棵棵树连根拔起！它这是为了争夺霸权，在与对手恶战前测试一下自己的实力。大象分为两种：印度大象和非洲大象。非洲大象的耳朵比印度大象大得多，把整个肩膀都遮住了，一直垂到腿上。大象是群居动物，每一群都有一个头象，一般都是最大最壮的那个。头象统治着整个象群，指导大象的行动，危险来临时发出信号。象鼻肌肉发达，对大象至关重要。著名的博物学家库利耶说，象鼻上的肌肉不少于四万块，每块肌肉都有其独特的作用，所以，象鼻的触觉和嗅觉特别灵敏，连地上的一根大头针它也能捡起来，摘树上最小的树叶也不在话下。大象一般十英尺高（约三米），有时候能长到十二英尺（约三米六）高，能活到七八十岁。

34

A Wise Dog
聪明的狗

There is a curly retriever at Arundel bearing the name of "Shock," which sets an example of good manners and intelligence to the animals which are not dumb. He carries the cat of the stables tenderly in his mouth, and would carry the kitten, but at present the kitten prefers its own means of locomotion. When Sanger's elephant got into trouble in the river Arun, this wise Shock was sent to turn him out, and his perseverance succeeded. He often will insist on carrying a bundle of umbrellas to the station, and safely he delivers them to their owners, and then, with many wags of his brown tail, he demands a halfpenny for his trouble. This halfpenny he carries to the nearest shop, lays it on the counter, and receives his biscuit in return. Need we say this dog has a kind, sensible master?

英国西阿克塞斯郡的阿伦德尔镇上有一只名叫"休克"的卷毛猎犬。它的礼貌和智慧为那些还不算笨的狗树立了榜样。在叼动马厩里的猫时它总是很轻很温柔,它也帮着挪动小猫崽,不过,现在小猫崽们更愿意自己行动。有一次,桑格家的大象在阿伦河里遇到了麻烦,聪明的"休克"被派去解难,它不屈不挠、顽强不懈地努力最终获得了成功。"休克"经常自告奋勇前往车站运送雨伞,把雨伞安全交到客户手里之后,它会摇起棕色的尾巴,索要半便士的辛苦费。然后,它会找一个最近的商店,把挣来的半便士放在柜台上,换取它心仪的饼干。还用我们说吗?它有个善良、睿智的主人。

35

Our Pet Hedgehog Timothy
我们的宠物刺猬蒂莫西

Timothy was our pet hedgehog. I bought him in Leadenhall Market, brought him home, and put him into the back-garden, which is walled in. There, to that extent, he had his liberty, and many, and many a time did I watch him from my study window walking about in the twilight among the grass, searching for worms and other insects. And very useful was he to the plants by so doing. When the dry weather came food got more scarce; then Timothy was fed with bread and milk from the back-kitchen window, which is on a level with the stone. Soon he came to know that when he was hungry there was the supply; and often he would come and scratch at the glass or at the back-door for his supper, and after getting it, walk off to the garden beds to make himself useful. Few people know of the great use of a hedgehog in a garden, or they would be more generally kept. Our Timothy, poor fellow, however, in spite of all his good qualities, came to a bad end. A strange dog coming one day, saw him walking about in search of his accustomed food, and pounced on him and bit him; still I had hopes of his recovery, but in a few days he died, and all of us were sorry.

蒂莫西是我在利登霍尔市场上买的宠物刺猬。我把它带回家，放在后花园里。花园四周都是墙，它可以在墙内自由活动。暮色朦胧中，透过书房的窗户，我无数次看着它在草地上徘徊穿梭找虫子吃，这对植物很有好处。随着旱季来临，虫子越来越稀少，于是，我们就把面包和牛奶放在后厨房的窗台上，因为花园里的石头刚好和窗台一样高。很快，它就知道要是饿了的话，那个地方会有吃的。所以，吃晚餐的时候，它会赶过来，挠挠玻璃或者后门；吃完之后，就回到花圃里继续做贡献。很少有人知道刺猬在花园里还能这么有用，否则的话，养刺猬会更普遍。可是，我们可怜的蒂莫西，那么乖的小家伙，结局却很惨。有一天，不知谁家的狗进来，看见它在那里找食吃，就扑过来，咬了它一口。当时，我还抱着一丝希望，盼着它能康复。可是，几天之后，它死了。我们大家都很难过。

36

The Brave Cockatoo
勇敢的美冠鹦鹉

One Charles Durand, of whose travels and adventures a book has been written, owned a cockatoo, which he carried about with him on his journeys; the bird's name was Billy, and he seems to have been as wise as he was loving. Charles was asleep in his tent, when he was roused by a sharp, shrill cry of the bird, of "Time to rise! time to rise!" accompanied by a violent flapping of the wings. So awakened, Charles looked around, wondering what had disturbed his feathered friend. The cause was soon plain—a deadly snake lay coiled up close to his bed, prepared to spring on the defenseless man. Just when he thought that all hope was at an end, the brave cockatoo sprang from his perch, seized the reptile by the neck, and held him tight till his master could summon help.

一个叫查尔斯·杜兰德的人有一只美冠鹦鹉,出门旅行时他总是带着它,还有一本专门写他冒险旅程的书呢。鹦鹉的名字叫比利,它不但忠心耿耿,而且聪明异常。"快起!快起!"那天,查尔斯在帐篷里睡得正香,突然鹦鹉的尖叫和翅膀猛烈的扑棱声惊醒了他。他不由得四下张望,不知什么惊扰了他的长羽毛朋友。原因很快就清楚了——一条能致人死命的毒蛇蜷在他的床前,正虎视眈眈地盯着他这个毫无防备的人,随时准备发起进攻。就在他以为一切都完了的时候,勇敢的美冠鹦鹉从栖木上猛扑了下来,一下子啄住了毒蛇的脖子,紧紧地咬住不放,直到主人找来帮手。

37

Hare Taking the Water
会凫水的兔子

I was pike-fishing one season on the Dorset Stour below Canford Major, when on passing from one field to another, I disturbed a hare. The animal at once entered an open, dry drain, and I lost sight of her. Presently, as I silently made my way plying my rod by the bank, I saw her, this time without any appearance of alarm, take to the water, and making her way through the sedges. She put her head to the stream so that the force of the current, with but slight exertion by swimming on her part, carried her nearly in a straight line to the opposite bank. Here I watched her to see whether she would trundle herself like a dog, but she merely rested a bit, letting the water run from her, and then set off at a rattling pace across the mead, which doubtless soon thoroughly dried her.

有一年，我在坎佛德·梅杰下面的多塞特郡斯陶尔河上钓梭鱼的时候，惊动了田里的一只野兔。小家伙一下子就蹦进了一条敞开着的干排水沟里，然后一眨眼就不见了。过了一会儿，我在河岸上静悄悄地收放鱼竿的时候，又看到了她。这次，她没有惊慌失措，而是穿过沙草跑到水边，一头扎进小溪，借着水势，轻轻划动身体，像一条直线似的游到了对岸。我站在岸这边，想看看她会不会像狗一样打滚，把身子弄干。可是，她只稍稍站了一会儿，让身上的水自然滴落，然后，撒开脚步，一路小跑着冲进了草丛。不用说，草丛很快就会彻底地把她身上的水弄干。

38

The Beaks of Birds

鸟嘴

I dare say you notice that all the birds in this picture have long beaks. We may be sure from this that they live in places and seek for their food in ways in which long beaks are just what they want. The fact is they are all marsh birds, and the soil of marshes being wet and soft, and full of worms, these long beaks enable them to probe it, and so get at the worms. I think the beaks of birds afford a striking example of how good God is in adapting creatures to the mode of life He has appointed for them. The eagles and hawks, you know, are provided with strong, short bills to enable them to seize and tear flesh. Those of canaries and all the finches are just the very instruments to crack seeds with. Parrots, with their tremendous weapons, can crush the hardest nuts of the tropic forest. The crossbill is fitted with a wonderful tool for tearing fir-cones to pieces. Robins and the other warblers have soft bills, which are all they want for eating insects and grubs.

你已经发现这张图里鸟的嘴巴都很长了吧？据此，我们能断定它们生活的地域和觅食的方式都离不开长嘴巴。事实也的确如此。这些鸟都是沼泽鸟。由于沼泽土壤又湿又软，里面各种软体虫子特别多，所以，鸟的长嘴巴可以伸进泥巴里，捉住虫子。上帝为各种生物选择了不同的生活方式，同时也为它们提供了适应这种生活方式的手段。我认为鸟嘴就是一个典型的例子，可以证明上帝考虑得是多么周全。鹰的嘴巴短而坚利，为的是能够猎取和撕咬猎物的肉。金丝雀等所有雀类的嘴巴都是天生的种子粉碎器。鹦鹉能用它们硕大无比的武器——嘴巴咬碎热带雨林里最硬的坚果。交喙鸟的嘴巴也很奇妙，能把杉果撕得粉碎。知更鸟和其他鸣鸟的嘴巴比较柔软，刚好适合吃昆虫和蛆虫。

39

Blackbirds and Young
大黑鸟和小黑鸟

A country lad having taken the nest of some blackbirds containing young ones, made off with it, but was closely pursued by the parents, who tried to peck his face so as to make him give them up. Mr. Jesse relates a similar instance, where a pair of old birds followed a boy into a house, pecking at his head while he was carrying off one of their young ones. People little think of the misery they cause when they rob the birds of their nestlings.

The bird's nest is thus described:

Now put together odds and ends,
Picked up from enemies and friends;
See bits of thread and bits of rag,
Just like a little rubbish bag.

一个乡村的小男孩摘了一个黑鸟窝,鸟窝里还有几只小黑鸟呢。他拿起鸟窝就跑,可是,大黑鸟却紧追不舍,一边追还一边啄他的脸,好让他就此罢休,放了小黑鸟。杰西先生也讲了一个类似的故事:有个男孩拿走了一只雏鸟,鸟爸爸和鸟妈妈一路跟到他家里,还不断地啄他的头。人们很少想到,夺走嗷嗷待哺的雏鸟会让鸟爸爸和鸟妈妈多么痛苦!

我们这样描述鸟窝的模样:

快把那零七八碎的东西攒起来,
管他是敌人的,还是朋友的。
瞧!那一根根线头,还有那一块块布片,
真像是一个小小的垃圾袋。

40

A Useful Pilot
管用的领头羊

There is a trained sheep kept on board a steamer plying in California. It goes out on the gang-plank, when a flock is to be loaded, to show that the approach is safe, and to act as pilot to the flock, which readily follows it on to the boat. The sheep, when in a flock, are all alike timid, and it is difficult to find a leader among them, each being afraid to go first; but when one goes, they all follow after, so that this clever sheep is very valuable. The only other way to get a flock on board a ship is to catch one and drag it on board; but this is not such a good way as having the clever "Pilot."

在往返于加利福尼亚的一艘汽船上，有一只训练有素的羊。每次要装运羊群的时候，这头羊就肩负起领头羊的作用，率先踏上跳板，以示此行安全，尽可仿效，羊群随即跟着它上船。一群羊在一起的时候，个个怯懦，谁也不愿意打前站，所以，很难在它们中间找到一只领头羊。但是，一旦有一只羊挺身而出，其他的羊也都会尾随其后，正因如此，这只聪明的羊弥足珍贵。除此之外，还有一个办法能让羊群上船，就是逮住一只羊把它拽上船；但是，这个方法不如养只聪明的"领头羊"高明。

41

A bear named Jack
一只名叫杰克的熊

The name of the bear is "Jack". I fetched him from the West India Import Dock on the 5th of November, 1870. He was running about with another bear on board ship, but the job was to catch him. After many attempts we at last put a strong collar round his neck, to which was attached a long chain, and then we got him into a large barrel, and fastened the head on with hoop-iron, lowered him over the side of the vessel into a boat, and then pulled to the quay, and hauled him up into a cart. For a time the little fellow was quiet enough, but he got very inquisitive when being driven toward the city, and wanted to have a look round. I managed to quiet him by giving him pieces of lump sugar. He arrived safely at the Crystal Palace, and has lived in an aviary till the beginning of last month, when he was put into his new bear-pit. The little fellow has grown twice the size he was when he first came. He is very playful, but sometimes shows his teeth when he is teased.

这只熊的名字叫"杰克"，是我在1870年11月5日那天从西印度进口码头上弄来的。当时，它正和另外一只熊在船上乱跑，我们的任务就是抓住它。我们试了好多次，才把一个拴着长链子的项圈结结实实地套在了它的脖子上。我们把它装进一个大桶里，用铁箍把桶盖箍好，然后把桶吊到船外的一条小船上，拉到了码头，接着，又连拖带拽地把它装到了一辆运货车上。刚开始，这个小家伙还比较安静，可是，车开往市里的时候，它不安分起来，总要东张西望。我给了它几块方糖，它才消停了下来。安全抵达水晶宫后，我们把它临时安顿在一个大鸟舍里面，直到上月初才把它接到专门为它修建的熊苑里面。现在，小家伙已经是刚来时的两倍大了。它很顽皮，可是如果有人捉弄它，它会生气地冲人家龇牙。

42

A Singular Habit of the Woodcock
丘鹬的奇特习性

Among several curious habits of the woodcock, described by the editor of the *Zoologist*, its practice of carrying its young is perhaps the most interesting. The testimony of many competent witnesses is cited to corroborate the statement. The late L. Lloyd, in his *Scandinavian Adventures*, wrote, "If, in shooting, you meet with a brood of woodcocks, and the young ones cannot fly, the old bird takes them separately between her feet, and flies from the dogs with a moaning cry."

The same author makes a similar statement in another work, this habit of the woodcock having been observed by a friend.

One of the brothers Stuart gives, in *Lays of the Deer Forest*, a graphic account of the performance. He says, "As the nests are laid on dry ground, and often at a distance from moisture, in the latter case, as soon as the young are hatched, the old bird will sometimes carry them in her claws to the nearest spring or green strip. In the same manner, when in danger, she will rescue those which she can lift; of this we have frequent opportunities for observation in Tarawa. Various times when the hounds, in beating the ground, have come upon a brood, we have seen the old bird rise with the young one in her claws and carry it fifty or a hundred yards away."

《动物学家》的编辑描述了丘鹬的一些奇特习性,其中最有趣的也许就是它搬运小宝宝的绝活了。这可是许多人的亲眼所见。已故的 L.利罗德在《斯堪的纳维亚历险记》一书中写道:打猎的时候,如果你碰到一窝丘鹬,里面还有不会飞的小丘鹬,大丘鹬就会用脚一个个地夹起小丘鹬,飞起来逃离猎狗的追捕,一边飞还一边凄厉地鸣叫。

这位作者在另一本书里也记录了类似的情形,是一个朋友的亲眼所见。

在《鹿林往事》一书中,斯图尔特俩兄弟中的一个对此也做了生动的描述。他说,"由于鸟巢筑在干燥的地上,离潮湿的地方有一段距离,所以,幼鸟一孵出来,大鸟就会用爪子夹着幼鸟飞到最近的泉水边或者绿地上。遇到危险,她也以同样的方式把能带走的幼鸟都带走。在塔拉瓦岛(西太平洋岛国基里巴斯的主岛)看到这种情形的机会比较多。很多时候,猎狗捕猎时会碰到一窝幼鸟,于是我们就看见大鸟用爪子夹着幼鸟飞起来,把幼鸟带到五十或者一百码之外的地方。"

43

The Sky-Lark
云雀

Has any one ever told you that they were "happy as a lark," and have you stopped to think how happy a lark is? —its joyous flight up into the sky, as high or higher than the sight of man can reach, singing louder and louder, and more and more gayly the higher it ascends? When the sweet hay-time comes on, and mowers are busy in the fields with their great scythes, it is sometimes a dangerous season for larks, who make their nests on the ground. Often the poor little nests must suffer; but only think how ingenious their owners are if they do. A mower once cut off the upper part of a lark's nest. The lark sitting in it was uninjured. The man was very sorry for what he had done; but there was no help for it—at least so he thought. The lark knew better, and soon afterward a beautiful dome was found made of grass over the nest by the patient, brave bird.

可曾有人对你说过他们"快乐得像云雀一样"？你又可曾停下来想过云雀到底有多快乐？它们越飞越高，歌声也越来越响亮、越来越欢快，它们飞到我们的目光所及之处，飞出我们的视野，那是多么快乐的飞翔！香甜的干草季节，人们挥舞起大镰刀在田间割草。对于把巢筑在地上的云雀来说，这个季节有时危机四伏。那些可怜的小鸟巢经常要遭殃；但是，如果真的遭殃的话，就想想鸟巢主人是多么有才吧！一次，有人割草时割掉了云雀窝的顶部，里面的云雀却毫发未伤。割草人为此感到很难过，可是却觉得事已至此，无能为力了——至少他是这样想的。云雀可聪明多了，困难面前毫不气馁、决不退缩，没过多久，一个漂亮的拱形草盖就完工了。

44

The Story of a Seal
海豹的故事

Some years ago a German Artist was traveling in Norway, on foot, with his knapsack on his back and his stick in his hand. He lodged most of time in the cottages that he fell in with on his road. In one of them there was a seal, which the fisherman had found on the sand, after harpooning the mother of the poor animal. No sooner was it admitted into the cottage than the seal became the friend of the family and the playmate of the children. It played from morning till night with them, would lick their hands, and call them with a gentle little cry, which is not unlike the human voice, and it would look at them tenderly with its large blue eyes, shaded by long black lashes. It almost always followed its master to fish, swimming around the boat and taking a great many fish, which it delivered to the fisherman without even giving them a bite. A dog could not have been more devoted, faithful, teachable, or even more intelligent.

几年前，一位德国艺术家背着背包，拿着手杖在挪威徒步旅行。多数时候他就在路边随便找个村舍投宿。其中有一家村舍里养着一头海豹，那头海豹是渔夫在沙滩上捡到的，当时，这个小可怜的妈妈已经丧命在他的鱼叉之下了。小海豹一进村舍就成了这家人的朋友和孩子们的玩伴。它和孩子们从早玩到晚，舔他们的手，细声细气地呼唤他们，声音和人没什么两样。黑黑的睫毛，蓝蓝的大眼睛，看着孩子们时总是那么温和。每次主人去捕鱼，小海豹几乎都跟着。它在船的四周游来游去，捉很多的鱼。它把这些鱼全都如数交给渔夫，连咬都不咬一口。即便是猎狗，也没有它敬业，没有它忠诚，没有它可教，甚至也没有它聪明。

45

The Bee
蜜蜂

OH! busy bee,
On wing so free,
Yet all in order true;
Each seems to know,
Both where to go,
And what it has to do.

'Mid summer heat,
The honey sweet,
It gathers while it may;
In tiny drops,
And never stops
To waste its time in play.

I hear it come,
I know its hum;

It flies from flower to flower;
And to its store,
A little more
It adds, each day and hour.

啊！忙碌的蜜蜂，
自由地飞，
有序地归；
个个似乎都知道，
要去哪儿，
做什么。

盛夏的炎热，
蜂蜜的香甜，
全力采集；
一小滴一小滴，
从不停歇
从不贪玩。

我听见它来了，
我熟悉它的嗡嗡声；
从这朵花飞到那朵花；
然后飞到蜂箱里，
再多一点
天天在贡献。

蜜蜂

46

Sheep Dog
牧羊犬

The dog that you see here looking quite maternal with her family around her, is the sheep dog, the shepherd's faithful and invaluable friend. It is the most sagacious and intelligent of all dogs, and volumes of anecdotes might be written of its intelligence and affection. Mr. St. John, in his *Highland Sports*, tells the following: "A shepherd once, to prove the quickness of his dog, who was lying before the fire where we were talking, said to me in the middle of a sentence concerning something else, 'I'm thinking, sir, the cow is in the potatoes;' when the dog, who appeared to be asleep, immediately jumped up, and leaping through the open window and on to the roof of the house, where he could get a view of the potato field, and not seeing the cow there, he looked into the farm-yard, where she was, and finding that all was right, returned to his old position before the fire."

瞧这只狗,膝下儿女成群、浑身散发着母爱的光辉。这是一只牧羊犬,既是牧羊人忠实的朋友,也是他们的无价之宝。在所有的狗中,牧羊犬最有悟性,最聪明。关于它的智慧和忠诚,有很多逸闻趣事,可以写成好几本书呢。圣·约翰先生在《高原运动》一文中就讲了这样一件事情:"一次,我们围坐在火边聊天,跟前躺着牧羊犬。为了证明牧羊犬反应灵敏,牧羊人说着说着突然冒了一句与我们谈话主题无关的话:'先生,我在想母牛跑到土豆地里去了。'话音刚落,似乎正在熟睡的牧羊犬一下子蹦了起来,从开着的窗户冲了出去,跃上屋顶,查看土豆地里的情况,土豆地里没有母牛;它又朝院子望去,原来母牛在那里,一切正常。于是,它又重新回到火边躺了下来。"

47

The Friendly Terns
有情有义的燕鸥

One day Mr. Edward, the Scotch naturalist shot at a tern, hoping to secure the beautiful creature as a specimen. The ball broke the bird's wing, and he fell screaming down to the water. His cries brought other terns to the rescue, and with pitiful screams they flew to the spot where the naturalist stood, while the tide drifted their wounded brother toward the shore. But before Mr. Edward could secure his prize, he observed, to his astonishment, that two of the terns had flown down to the water, and were gently lifting up their suffering companion, one taking hold of either wing. But their burden was rather heavy; so, after carrying it seaward about six or seven yards, they let it down, and two more came, picked it up, and carried it a little farther. By means of thus relieving each other they managed to reach a rock where they concluded they would be safe.

一天，苏格兰自然学家爱德华先生射中了一只燕鸥，想要把这只美丽的鸟制成标本。子弹击穿了燕鸥的翅膀，燕鸥尖叫着掉到了水里。它的哀鸣引来其他燕鸥前来援救，它们凄楚地尖叫着，飞到了自然学家站着的地方。这时，海潮已经把它们受了伤的弟兄冲到了岸边。可是，爱德华先生还没来得及捡起他的战利品，就看到了令他吃惊的一幕。两只燕鸥飞到了水边，一边一个叼住伤者的翅膀，轻轻地抬起了在痛苦中挣扎的同伴。可是，负担太重了，他们向海边飞了大约六七码，就把伤员放了下来。这时，又来了两只燕鸥，重新叼起伤员，又往前飞了一段。就这样，它们轮流替换着，飞到了一块岩石上。对它们来说，到了这里就安全了。

48

The Otter
水獭

The otter belongs to a class of animals which we may call the Weasel tribe. Their bodies are long and lithe, and their legs short. This family includes the weasel (its smallest member), the stoat, the ferret, the pole-cat, the marten, and the otter (its largest member). You may then think of the Otter as a water-ferret, or water-weasel. He can swim most elegantly, and he is a beautiful diver. Let a fish glide underneath him, and he is after it in a moment; and as the fish darts here and there to escape, the Otter follows each rapid movement with unerring precision. When the fish is caught, the Otter carries it to the bank and makes a meal. But the Otter is like naughty Jack who leaves a saucy plate—he spoils much more fish than he eats. The trout and other fish are so much alarmed at the appearance of an Otter that they will sometimes fling themselves on the bank to get out of his way.

水獭属于一个可称之为黄鼠狼部落的动物家族。它们体长而柔韧、四肢短小。这个家族里有黄鼠狼（数量最小）、白鼬、雪貂、鸡貂、貂鼠和水獭（数量最多）。你会以为水獭就是水雪貂，或者水黄鼠狼，因为它们不仅是游泳健将，而且还是潜水能手。鱼儿尽可以从它身边悄悄溜过去，但它旋即就会尾随而至；鱼儿左冲右撞想法逃命，它也忽左忽右紧追不舍，一步不落。捉住鱼之后，它就把鱼叼到岸边，美餐一顿。可是，水獭像顽皮的杰克一样，总是剩饭。它浪费掉的鱼比吃掉的还要多。鲑鱼和其他的鱼一看见水獭就惊慌失措，有时候，它们会自己蹦到岸上，为它让路。

The Mastiff

The mastiff is a large, grave, sullen-looking dog, with a wide chest, noble head, long switch tail, bright eyes, and a loud, deep voice. Of all dogs this is the most vigilant watcher over the property of his master, and nothing can tempt him to betray the confidence reposed in him. Notwithstanding his commanding appearance, and the strictness with which he guards the property of his master, the mastiff is possessed of great mildness of character, and is very grateful for any favors bestowed upon him. I once went into the barn of a friend where there was a mastiff chained; I went up to the dog and patted him on the head, when out rushed the groom from the stable exclaiming, "Come away, sir! He's dangerous with strangers." But I did not remove my hand nor show any fear. The consequence was, that the dog and I were the best of friends; but had I shown any fear, and hastily removed my hand, I might have fared rather badly, for this dog always couples fear with guilt.

　　獒是一种大型犬。它身体高大、尾巴很长、昂头阔胸、目光炯炯、声音浑厚有力。獒不苟言笑,总是一副闷闷不乐的样子。在所有的犬中,獒是主人财产最警觉的守护者,无论什么都不会诱使它背叛主人对它的信任。獒尽管外表吓人,而且履行起职责来不留情面,但是,却天性温和、知恩图报。有一次,我在一个朋友的牲口棚里看到一只用链条拴着的獒。于是,我走上前去拍了拍它的脑袋,可就在这时,马夫从马厩里冲出来大喊:"走开,先生!危险!"但是,我并没有把手缩回来,也没有惊慌失措。结果,我俩成了最好的朋友。当时,如果我流露出哪怕一点点的恐惧,迅速缩回伸出去的手,那我可能就惨了,因为獒总是把恐惧当成是心怀不善。

50

The Cunning Wood–Pigeons
狡黠的木鸽

One who loves our feathered friends has described a curious instance of their instinct. On the back lawn at a gentleman's house, they have a feeding-box for the pheasants, which opens on their perching upon it, but remains shut if any lesser bird than a hen pheasant perches there, which saves the contents from the thefts of these, and of rats, mice, and other vermin. But the gentleman discovered that the contents of the box was being more rapidly emptied than the wants of the pheasants warranted. So he kept a watch on the box, and soon discovered a wood-pigeon perch on the box, but his weight not being sufficient to open the lid, he beckoned to another pigeon, and their combined weight made the lid fly open, and after each had taken what they required, they flew away, and the box closed with a "click."

我们的羽类朋友也有急中生智的时候。有个爱鸟人士就讲了这样一个奇妙的故事。一位先生在自家房后的草坪上放了一个给雉鸡喂食的盒子，雉鸡落在盒上可以靠身体重量压开盒盖，但比它轻的鸟落上去则无法打开食盒。因此，其他鸟，还有大小老鼠和别的坏家伙就无法偷吃盒子里的食物了。但是，这位先生发现盒子里的食物越来越供不应求了。于是，他开始在暗中观察。没过多久，就看见一只木鸽落到了食盒上。不过，它的身体太轻，无法打开盒盖，于是，它唤来另外一只木鸽，齐心协力压开了盒盖。各取所需之后，两只木鸽飞走了，而盒盖则"咔嗒"一声又合上了。

51

Sea Reptiles
海洋爬行动物

There were in the sea in very ancient times—long before the flood—two very large and wonderful reptiles. Of them we present striking illustrations. One of them has been named the Ichthyosaurus, which means Fish Reptile. Its head somewhat resembled that of the crocodile, except that the orbit was much larger, and had the nostril placed close to it, as in the whale, and not near the end of the snout. It had four paddles and a powerful tail, and was very active in its movements and a rapid swimmer.

The other huge reptile was the Plesiosaurus, the meaning of which is "Near to a Reptile". Its structure was very singular and its character very strange. In the words of Buckland: "To the head of a lizard, it united the teeth of the crocodile, a neck of enormous length, resembling the body of a serpent, a trunk and a tail of the size of an ordinary quadruped, the ribs of a chameleon, and the paddles of a whale".

在还没发生大洪水以前很久的远古时代,海里有两个神奇的巨型爬行动物,我们把它们画了下来,这幅插图就是。这两个爬行动物中的一个叫鱼龙,就是像鱼一样的爬行动物。鱼龙的头有点像鳄鱼头,不过眼眶比鳄鱼的大。它的鼻孔不在吻部末端,而是靠近眼眶,和鲸鱼一样。它有四个桨鳍和一条有力的尾巴,行动灵活自如,是个游泳健将。

另一个巨型爬行动物蛇颈龙就是"接近爬行动物"的意思。蛇颈龙的身体结构独特、性情古怪。用英国著名地质学家巴克兰的话说就是:"蜥蜴的脑袋鳄鱼的牙,脖子奇长像蛇身;四足动物的鼻子和尾巴,变色龙的脊椎、鲸鱼的鳍。"

52

Swiss Mountain Scenery
瑞士山景

In Switzerland, one of the chief employments of the people is that of herdsmen and shepherds, and nearly the half of the surface of the country is occupied as mountain pastures and meadows. Here you see the woman tending the sheep and goats, and spinning industriously, while her husband is busy with some other part of the duties of tending the sheep. It is often painful to see how much the poor sheep and oxen suffer while being driven through the streets. It is pitiful to see them looking in vain for some place of rest and shelter. Little boys in towns sometimes like to HELP—as they call it—to drive cattle, but they generally increase the terror and confusion of the poor beasts, and little think of the pain they are causing. Sheep and goats are very useful to us; besides serving us for food, they supply our cloth and flannel clothes, blankets, and other warm coverings.

在瑞士，人们从事的主要职业之一就是牧羊。全国几乎一半的土地都用作了山地牧场和草场。在这里，你能看到妇女们辛勤地照看绵羊和山羊，并且纺线织布，而丈夫们也为了羊而劳碌奔波，不辞辛苦。可怜的绵羊和牛群被驱赶着穿街过巷，那么痛苦不堪，看着真让人难过！它们茫然四顾，多么想找个地方歇歇脚，可是，却不能够，看着真让人心碎！有时候，镇上的小男孩自告奋勇来赶牲口——他们说想"帮忙"，可是，往往只会帮倒忙，让可怜的牲畜更加惊慌、不知所措，他们哪里想得到自己给这些牲畜带来的痛苦。绵羊和山羊，都是我们的宝贝，它们不仅供我们吃，还供我们穿、供我们盖，那些布匹、法兰绒衣服、毯子和其他保暖衣物都是它们的功劳。

53

Partridge and Young
大鹑鹑和小鹑鹑

One afternoon, while walking across a meadow, near a village, I saw a dog of the terrier breed pursuing a partridge, which every now and then turned and made at it with its wings down, then rolled over, then ran, and again rushed at the dog. I drove the dog away, when I was surprised to see a number of young partridges running from behind the old bird who had been trying to protect them from the dog, and guarding their retreat. So you see how brave the most timid creatures become when in danger, and when their young are near. Instinct tells them that they have to protect their little ones, and risk everything, even their own lives, for their safety. We can get beautiful lessons every day from the birds and poor dumb animals, if we only study them as we ought.

一天下午,我穿过一个村庄附近的草地时,看到一只小猎犬追赶一只鹌鹑。鹌鹑不时地回过头来,扑棱起翅膀迎击小狗,然后,转过身子接着跑。过一会儿,又回过头来迎击小狗。我赶走了小狗,却惊奇地发现,从大鹌鹑身后跑出来好几只小鹌鹑。原来,刚才大鹌鹑一直在掩护这几只小鹌鹑撤退。所以,你瞧,为了保护身边的孩子,连最胆小的鸟儿遇到危险时,也会变得这么勇敢!本能要求它们必须保护自己的孩子,为了孩子们的安全,它们要敢于冒一切风险,哪怕是献出自己的生命。让我们关注身边的鸟儿和可怜的动物们吧!它们虽然不会说话,但是,只要我们留心观察,每天都会得到美好的启迪。

54

The Kingfishers' Home
翠鸟的家

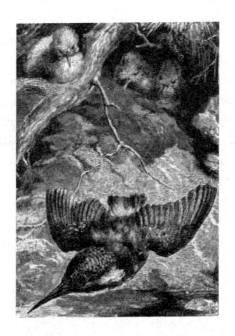

Very pretty birds were Mr. and Mrs. Kingfisher, with dark, glossy, green wings, spotted with light blue. Their tails were also light blue, and there was a patch of yellow near their heads. The little Kingfishers were quite as pretty as their parents, and Mr. and Mrs. Kingfisher were exceedingly proud of them.

"Only they eat a great deal," said Mr. Kingfisher; "I am getting very tired."

For Mr. Kingfisher had been flying backward and forward all day, and it was surprising to see the quantity of fish he caught for his family.

When he built his nest he took care that it should be near a stream, and he found one close by a high cliff that Mrs. Kingfisher said would be just the place; so they scooped out a deep hole, and there the eggs were laid, and in due time six little Kingfishers burst out of the shells.

翠鸟夫妇长得非常漂亮，深绿色的翅膀闪闪发光，上面还点缀着淡蓝色的小点；尾巴也是淡蓝色的，头下面还有一小块黄色。小翠鸟长得和爸爸妈妈一样漂亮，所以，翠鸟夫妇别提有多自豪了。

不过翠鸟先生说，"美中不足就是孩子们吃得太多了，让我疲惫不堪。"

为了养家，翠鸟先生整天都飞进飞出去捕鱼，它捕的鱼数量惊人。

翠鸟爸爸筑巢的时候，尽量选择靠近河流的地方。它在一个高高的悬崖附近找到了这样一个地方，翠鸟夫人也说这个地方正合适。于是，它们挖了一个深洞，在那里产了蛋，不久，六只小翠鸟就如期破壳而出了。

55

Rats Carrying Eggs Upstairs
搬鸡蛋上楼的老鼠

Rats are very ingenious little creatures; they have actually been known to convey eggs up a staircase, from the pantry to their nest! Here is a beautiful picture, by Mr. Harrison Weir, from *The Children's Friend*, showing how they did it. The rat bears little resemblance to the rats with which we are chiefly acquainted, namely, the black rat, the albino or white rat, and the brown rat.

The other day, as I was walking by the river-side, I saw a beautiful little creature sitting on a stone in the stream, with a piece of succulent root between its forepaws, and nibbling its repast in perfect peace with every living thing. It was timid and innocent in the expression of its countenance. Its color was of a reddish brown. It was about as large as the common rat of the sewers, but its tail was much shorter, and covered with hair.

老鼠是非常机灵的小生物。据说它们曾经把鸡蛋从楼下的食品储藏室搬到了楼上的窝里！它们是怎么搬的，看这幅图就知道了。这是哈里森·威尔先生为《孩子们的朋友》画的插图。图上的老鼠和我们平时熟悉的那些黑鼠、白化鼠或白鼠以及棕鼠不太一样。

有一天，我在河边散步的时候，看到一只漂亮的小老鼠坐在河流中间的一块石头上，前爪捧着一块汁多味美的菜根在那里啃得忘乎所以，一副胆怯、天真的表情。它身体呈红棕色，大小和下水道里常见的老鼠相差无几，只是，尾巴短多了，还毛茸茸的。

56

Heron
苍鹭

The heron when attacked by an eagle or falcon endeavors to escape by rising in the air and getting above its foe. The wings of the heron strike the air with an equal and regular motion which raises its body to such an elevation that at a distance nothing is seen except the wings, which are at last lost sight of in the region of the clouds. If its enemy gets above it, and upon or near its body, it defends itself vigorously with its long and powerful beak, and often comes off victorious. The heron frequents the neighborhood of rivers and lakes. Almost always solitary, it remains for hours motionless on the same spot. When seeking the fish or frogs on which it chiefly feeds, the heron wades into the water, folds its long neck partially over its back and forward again, and with watchful eye waits till a fish comes within reach of its beak, when it darts its head into the water and secures its slimy, slippery prey.

苍鹭遭到鹰的攻击时，逃命之计就是飞到天上，超越敌人。但见它均匀有力地拍打着翅膀，越飞越高，开始还能看清它的翅膀，后来就完全消失在云端，什么也看不见了。如果敌人在它的头顶盘旋，或者扑到近前，它就用厉害的长嘴全力自卫，往往可以成功脱险。苍鹭经常出没于河流和湖泊四周，它总是形单影只，能在原地一动不动地站好几个小时。苍鹭主要吃鱼和青蛙。为了捕食，它站在水中，长长的脖子一会儿弯到后面，一会儿又伸到前面，警惕地注视着水面，一旦有鱼进入它嘴巴够得着的地方，它就一头扎进水里逮住那溜光水滑的猎物。

57

A Horse Guardian
守护主人的马

On one occasion a gentleman was returning home from a fatiguing journey, and became very drowsy. He fell asleep, and, strange to say, he also fell from his saddle, but in so easy a manner that the tumble did not rouse him, and lay sleeping on where he alighted. His faithful steed, on being eased of his burden, instead of scampering home as one might have expected, stood by his prostrate master, and kept a strict watch over him. Some laborers at sunrise found him very contentedly snoozing on a heap of stones. They wished to approach the gentleman, that they might awaken him, but every attempt on their part was resolutely opposed by the grinning teeth and ready heels of his determined and faithful guardian. They called out loudly, and the gentleman awoke and was very much surprised at his position, while his faithful horse showed his pleasure by neighing and scraping his feet on the ground. The gentleman then mounted, and they galloped away at great speed, both glad to be able to make up for lost time.

有一次,一位先生结束了疲惫不堪的旅行,骑着马返程回家。途中,他太困了,不知不觉地打起了盹,从马鞍上摔了下来。可是,说来也怪,这一跤竟然没有把他摔醒,相反,他却躺在那里接着睡了起来。他那忠诚的坐骑,一下子没了负担,可出乎大家意料的是,它并没有撒开四蹄疾跑回家,而是待在席地而卧的主人身边,认真地站起岗来。黎明时分,一些干活的人发现他在一堆石头上呼呼大睡,就想走过去叫醒他,可是,一切的努力都遭到了这个意志坚定、忠心耿耿的守护者的坚决阻挠,它冲着干活的人又是龇牙咧嘴,又是蹬蹄甩腿,弄得他们不敢靠前。于是,人们大声喊叫,才把这位先生唤醒。看到自己躺在地上,他大吃一惊,而他那忠诚的坐骑则高兴地嘶叫起来,不断地在地上蹭着蹄子。这位先生随即重新上马,嘚、嘚、嘚地飞驰而去,主仆俩都很高兴能把损失的时间赶回来。

58

Battle between a Fox and a Swan
狐狸和天鹅之战

A fierce battle between a fox and a swan took place at Sherborne Park. Master Reynard seems to have caught the old swan napping, and to have seized him by the throat. The bird defended himself with his wings so powerfully that its assailant was done to death in no time, and a workman going past the lake above the bridge next morning found both fox and swan lying dead together. The bird had received a fatal bite in the throat; the fox had one leg broken and the side of its head completely broken in. The swan was the oldest bird on the lake.

谢尔本公园里发生了一场狐狸与天鹅之间的恶战。事情好像是这样的：列那狐①看到老天鹅在打盹，就趁机过去咬住了它的喉咙。天鹅挥起翅膀奋力还击，顷刻间，就将来犯之敌置于死地。第二天早晨，一个从湖上小桥路过的工作人员看到狐狸和天鹅躺在那里，都死了。天鹅的喉咙被咬了致命的一口，而狐狸则断了一条腿，头上有一侧也完全凹陷了下去。这只天鹅是这个湖上最年长的鸟了。

【注】
① 民间故事中，常把狐狸称作列那狐。

59

Mutual Affection
友情

We have a beautiful long-haired little dog called Tousy, which lately had a pup. This queer little bantling was jumping and tumbling about the green one day, when a lady entered followed by a dog. Tousy made a ferocious assault on the four-footed stranger, by way of defending her young, and our magnificent white cat, which was sitting on the doorstep, seeing or supposing that his friend Tousy was in danger, made two immense bounds, and alighted on the back of the intruder, whose eyes would have been scratched out but for prompt rescue.

The mutual affection of these two animals is unbounded, and yet we hear human disagreements compared to cat-and-dog life! These animals, and many others, are capable of the most devoted affection to their young, and to their mates, and frequently teach us lessons of kindness to one another.

我家有一只长毛小狗叫图西，长得很好看。最近，它生了一只小狗崽。有一天，可爱的小狗宝宝在草地上跳来滚去玩得正开心，这个时候，有位女士进来了，后面还跟着一只狗。图西猛地朝那个四条腿的访客冲过去，以为人家会伤害它的小宝宝呢。当时，我家那只了不起的白猫正蹲在台阶上，它看到，或者说它以为自己的朋友遇到了危险，于是，大跳两下，落到这位不速之客的背上。如果不是及时相救，这位不速之客的眼睛早就被猫抠出来了。

猫和狗之间有着深厚的友情。可是，人们竟然经常把相互之间的争执比作猫狗之争，简直是岂有此理！这些动物以及其他很多动物，对它们的孩子和伙伴同样充满了无私的爱，而且，常常让我们受到爱的教育。